SEA CREATURES AND SEA SHORES

Peter Day
Antony King
James Manna

An underwater guide to the Gulfs of South Australia

Peter R Day Resource Strategies
PO Box 271
Blackwood, SA 5051
www.resourcestrats.com.au

ISBN: 978-0-9872125-0-4

Text:

- Peter Day – Peter R. Day Resource Strategies – www.resourcestrats.com.au

Photographs:

- Peter Day – Flinders University Underwater Club
- Antony King – Flinders University Underwater Club
- James Manna – Flinders University Underwater Club

Acknowledgements:

Technical reviewers have volunteered their time, amazing expertise and warm support. Heartfelt thanks to:

- Associate Professor Sean Connell – University of Adelaide
- Associate Professor Graham Edgar – University of Tasmania
- Thierry Laperousaz – SA Museum
- Dr Carl Mooney – Flinders University

Editing, design and layout:

- Anne Burgi – SUBStitution Pty Ltd – www.substitution.com.au

Sea Creatures website:

- www.seacreatures.net.au

CONTENTS

ABOUT THIS BOOK

This book is a handy guide for divers, snorkellers and beachcombers.

There is a wonderful underwater world to be discovered at Adelaide's doorstep. This book opens a door to that world and the marine life that lies hidden just off the shore in South Australia's Spencer Gulf and Gulf St Vincent. It illustrates commonly observed creatures and makes it easy for people to identify them, and it explains the relationships between marine lifeforms and their environment.

This book will help snorknellers, scuba divers and beachcombers identify the marine creatures they find, and understand their unusual characteristics and how they fit into the ecology of the Gulfs. It unveils the evolution of marine organisms and their physiology and social interactions. It is an introduction to that 'other world' for those interested in exploration, but it does not aim to be a comprehensive reference book. Others are better at that (see listed references such as Coleman, Edgar, and Gowlett-Holmes).

The focus is on the Gulf St Vincent and Spencer Gulf, which are close to Adelaide and offer some great diving and snorkelling spots. They host a wide array of marine organisms and are home to some creatures that cannot be found anywhere else in the world, although it is also relevant to much of southern Australia.

We begin with an explanation of the environments that harbour marine life and make interesting dive and snorkelling sites: reefs and wrecks, jetties, seagrass meadows and rocky shorelines. Each is discussed in turn, with descriptions of popular dive and snorkelling sites.

Many marine life forms do not have 'common' names and are difficult to identify. Therefore, some scientific names and classifications are introduced. This is done by outlining how different groups have evolved. Having introduced the main groups of marine plants and animals it is then possible to discuss each in more detail and to look at some of the species most commonly seen in the Gulfs.

There are tips at the end of the book for those contemplating diving or snorkelling for the first time.

This book provides advice on where to go for good diving and snorkelling experiences, describes and illustrates the creatures you will see underwater, and presents some information on how different species function and interact. It will be a handy guide for novice and experienced divers and snorkellers alike, and for anyone combing the shoreline. It will help readers appreciate the wonderful creatures to be encountered underwater and make exploring that world a richer experience.

This book will help people understand the creatures they see underwater and on the seashore – and their intriguing relationships with each other and their environment.

THE GULFS OF SOUTH AUSTRALIA
Port Augusta
Point Lowly
Whyalla
Peterborough
Port Pirie
Crystal Brook
Port Broughton
Cowell
Burra
Clare
Kadina
Moonta
Port Hughes
SPENCER GULF
Port Wakefield
Maitland
Ardrossan
Kapunda
Nuriootpa
Wardang Island
& Port Victoria
Gawler
Barker Rocks
Minlaton
GULF ST VINCENT
Stansbury
Adelaide
Mannum
South Australian
Yorketown
Edithburgh
Seacliff Reef
Hallett Cove
Point
Gilbert
Murray Bridge
Marion Bay
Port Noarlunga
Aldinga Reef
Strathalbyn
Tailem Bend
Haystacks Island
Myponga Beach
Ex HMAS Hobart
INVESTIGATOR STRAIGHT
Second Valley
Rapid Bay
Victor Harbor
Kingscote
BACKSTAIRS PASSAGE

THE GULFS

FORMATION OF THE GULFS

Geologically ancient fault lines and basins provide the foundations for the Gulfs, but their 'modern' history, the past million years or so, is dominated by changes in climate and sea levels. Over those million years, there have been repeated cycles of global cooling and warming which have resulted in the rise and fall of sea levels – repeatedly inundating and exposing the Gulfs.

The last 'interglacial warm period', when sea levels were similar to those of today, was around 125,000 years ago. Sea levels steadily declined from there until as recently as 15,000 years ago when the shoreline lay south of Kangaroo Island on the edge of the continental shelf and much of the earth's water was locked in ice. Rising at an average of 24 mm/year, the sea inundated the plains and salt lakes of the Gulfs to reach the current sea level 6,000-7,000 years ago. Stranded shoreline features and extensive areas of marine silts and salt marsh (samphires) provide some evidence that seas rose to be slightly higher than today (around 2 metres higher).

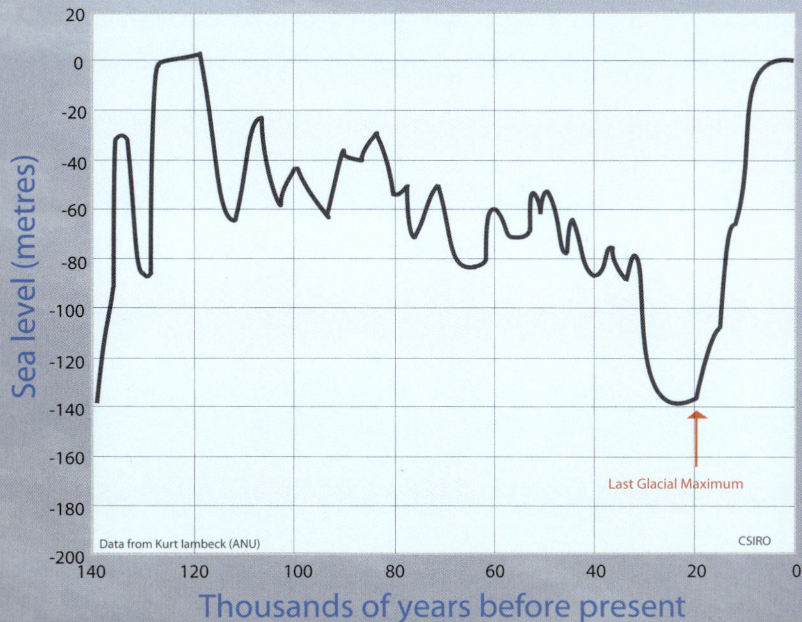

Figure 1: Sea level history over past 140,000 years (CSIRO, 2010)

TIDES, TEMPERATURES AND CURRENTS

South Australia's coast is influenced by two main ocean currents. The eastward flowing Leeuwin Current brings warm tropical water from north-west Western Australia to the west of the Great Australian Bight, while the Flinders Current flows westward from Victoria to the eastern portions of the Great Australian Bight. In some locations it is associated with the upwelling of cold, nutrient rich water from beyond the continental shelf during summer and autumn. Within the gulf region, upwelling occurs on the south western portion of Yorke Peninsula. The different currents have contrasting plants, animals and ecologies. South Australian coasts have many similarities with those in Western Australia.

In both Gulfs, currents tend to circulate clockwise from the south west, but cut-off eddies also form, e.g. in the upper (northern) reaches and off-shore from Myponga. The northern portion of Spencer Gulf (above Point Lowly) has limited connection with the rest of the Gulf. Tides generally move in a north – south direction and are important in flushing water in and out of the Gulfs. Dodge tides (when there is very little, if any, tidal movement in a day) are a phenomenon of the Gulfs, and are unusual around the world.

The Gulfs are warmer and more saline in the north and are considered to be 'seasonally subtropical systems'. They support several species that are 'relics' from previous warmer climates and are now only found in the Gulfs – and in subtropical waters far to the north on the eastern or western coast of Australia. Average water temperatures vary with the seasons, ranging from the low teens in winter to the mid-20s in summer.

Figure 2: Simulated circulation in Gulf St Vincent (Twidale et al, 1976)

The diverse marine conditions found in South Australia support an equally diverse range of plants and animals. Lying between the warm western and cooler eastern waters and having zones of cold upwelling, together with a range of landforms and marine habitats, South Australia has one of the most diverse marine floras in the world. Within the Gulfs, diversity is greatest along south west Yorke Peninsula (with more than 100 dominant species of seaweed) and least in the upper regions of the Gulfs (with 15-20 species of dominant seaweeds).

The relative isolation of the Gulfs is another reason for the diversity of species found there. Some species have been separated from their kind and have evolved differently to their relations in other parts of the world. The gulfs have a high degree of 'endemism', i.e. a relatively high number of species that are only found in those waters.

ENDEMISM: A measure of the number of species that are confined to a given region and not found anywhere else in the world.

References

Connell SD & Gillanders BM (2008) 'Marine Ecology'. Oxford University Press. South Melbourne. Victoria
– Chapter 9. Suthers IM & Waite AM. 'Coastal Oceanography and Ecology'

CSIRO Coastal Marine and Atmospheric Research website, Sea Level Rise. www.cmar.csiro.au/sealevel/sl_hist_intro.html

Bullock DA (1975) 'The general water circulation of Spencer Gulf, South Australia, in the period February to May'. Transactions of the Royal Society of South Australia. 99 (1) www.samuseum.sa.gov.au/Journals/TRSSA/TRSSA_V099/TRSSA_V099_p043p053.pdf

Shepherd SA, Bryars S, Kirkegaard IR, Harbison P, Jennings JT. (2008) 'Natural History of Gulf St Vincent'. Royal Society of South Australia Inc.
– Chapter 3. Fuller MK & Gostin VA. 'Recent coarse biogenic sediments in Gulf St Vincent'

– Chapter 4. Binnie MN & Cann JH. 'The distribution of Foraminifera in Gulf St Vincent'

– Chapter 19. Edyvane KS. 'Macroalgal biogeography and assemblages of Gulf St Vincent'

Twidale CR, Tyler MJ & Webb BP. (1976) 'Natural History of the Adelaide Region'. Royal Society of South Australia Inc.
– Chapter 11. Bye JAT. 'Physical oceanography of Gulf St Vincent and Investigator Straight'

Twidale CR, Tyler MJ, Davies M. (1985) 'Natural History of Eyre Peninsula' Royal Society of South Australia. Adelaide.
– Chapter 14. Glover CJM & Olsen AM. 'Fish and major fisheries'

Ward TM, Goldsworthy SD, Rogers PJ, Page B, McLeay LJ, Dimmlich WF, Baylis AMM, Wiebkin A, Roberts M, Daly K, Caines R & Huveneers C. (2008) 'Ecological importance of small pelagic fishes in the Flinders Current System.' Report to the Department of the Environment, Water, Heritage and the Arts. Report 276. SARDI Publication F2007/001194-1. www.environment.gov.au/coasts/mbp/publications/south-west/sw-flinders-pelagic-fishes.html

What ecology means

Ecology is the study of living things and their environment. It considers the distribution and abundance of species – where they live and in what numbers – and their interactions with each other and the environment. Surveys and maps are starting points for ecologists, but the real story is why different species occur where they do and what causes their distribution and abundance to change. Understanding the ecology of a species relies upon knowledge of:

- Life cycles – how they breed and what affects their life span
- Growth patterns – how they feed or grow, what they feed upon and what feeds on them
- Behaviour – how they interact with one another and other species
- Mobility – how they move and disperse to other sites.

These factors are discussed in the following chapters. The physical environment, the interactions between different species (e.g. predator – prey relationships), and the adaptation of species to these features are driving forces affecting where they occur and how well they survive.

Physical features important to the ecology of marine species include:

- water temperature
- light
- currents, tides and turbulence
- nutrient and sediment concentrations
- salinity
- the nature of the seabed, whether it is a sand, silt or rocky sea floor.

Four environments are considered in this book:

- rocky reefs and wrecks
- jetties
- seagrass meadows
- rocky shorelines.

1. Rocky reef

2. Seagrass meadow

3. Jetty

4. Rocky shoreline

Rocky reefs and wrecks

Ecology of reefs and wrecks

ALGAE: Aquatic organisms powered by the sun. Algae can fix the sun's energy and turn it into organic compounds in a process (photosynthesis) that also releases oxygen. They range from microscopic singe cells to large, multicellular organisms.

INVERTEBRATES: Animals without backbones, such as worms, crabs, corals, sponges and snails as well as insects and spiders. Invertebrates are found in the sea and on land, and make up more than 95% of all animal species (estimated to be 97-99%).

Schools of fish and the gentle rhythmic dance of swaying of seaweed typify a reef 'landscape' in the South Australian Gulfs. They are enticing places to explore. This section considers the ecology of the fish and algae associated with reefs and wrecks. Wrecks have many features in common with reefs, although many also have attributes very like jetties, which are discussed in the next section. The jetties chapter focuses on invertebrates, many of which can also be found on reefs and wrecks.

The distribution of many algae, including seaweeds, is strongly influenced by light. Light of different wavelengths (colour) penetrates to different depths and algae of different colour absorb light of different wavelengths. Algae grow best in conditions where their colour is a good fit with the wavelength of incoming light. Green algae tend to occur in shallower waters with more light penetration, brown algae in deeper water and red algae in areas of limited light.

Fish ecology is influenced by water temperature, currents, depth, the availability of food sources and habitat (such as crevices and holes), as well as predator – prey relationships. Seaweeds and other algae provide food for herbivorous fish, and shelter for herbivores and carnivores alike, so their diversity and abundance is an important factor in the distribution of fish species.

Many marine animals filter microscopic food from the water without discriminating between plants and animals, but others do. Animals can be described as:

Herbivores: Organisms that eat plants.

Carnivores: Organisms that eat other animals.

Omnivores: Organisms that eat plants and animals.

Planktivores: Organisms that eat plankton.

Detritivores: Organisms that eat decaying organic matter (detritus).

Fishes and seaweeds

Reefs and wrecks provide a range of environments supplying food, habitat and shelter. Each niche is attractive to different species of fish and seaweed. Around reefs, different species of fish may be found:

- in mid-water, above and around the reef
- above or within seaweed on the reef
- in caves and cavities within reefs
- within surrounding seagrass.

FISH AND FISHES: The term 'fish' is used for any single species of fish. 'Fishes' refers to two or more species of fish.

Fish encountered in mid-water are often carnivores, feeding on smaller fish, invertebrates or plankton. Many mid-water reef fish (e.g. snapper) may range widely, so their presence may be seasonal or purely a matter of chance. Some species, e.g. sweep, are omnivores – they often school near reefs and dine on plankton concentrated there by currents, but they will also feed on algae and on other animals.

Other fish are regularly found within a few metres of weedy reefs, where they feed either on seaweed or on animals living in association with the algae. Examples are:

- Wrasse (or 'rock cod') use a 'suction-bite' technique when eating, biting off algae and sucking small animals (e.g. crustaceans, molluscs and worms) from the seaweed.
- Dusky morwong, silver drummer and herring cale are examples of reef-dwelling herbivores. Many herbivorous fish rely on microbes in their gut to break down the vegetation and aid digestion.
- Zebrafish are omnivores, mainly because they ingest small crustaceans along with their primary food, algae.

PLANKTON: Tiny animals and plants that float freely in the water and are an important food source for many marine animals, both large and small.

Some species, usually with limited home ranges, spend most of their time under the algal canopy.

Fish that live in caves and crevasses are often planktivores or carnivores. Examples are:

- Bullseyes are nocturnal – coming out at night to forage on plankton.
- Moonlighters and western talma, which are carnivorous and pick tiny crustaceans from the surface of algae.
- Southern blue devils, colourful carnivores that shelter in caves and crevices.

Seagrasses that surround reefs host species that often feed on invertebrates such as worms, molluscs and crabs. Goatfish are an example. They feed on exposed sandy bottoms and use the sensory barbels under their chin to detect tiny crustaceans, which they suck up.

Some creatures specialise in removing external parasites from fish. The 'cleaning stations' they establish are usually near a prominent bottom feature where 'client' fish will come and adopt a cleaning pose – fins erect and mouth open. Cleaner fish (such as moonlighters and pencil weed whiting) remove parasites from the body, mouth and gills of their clients, resulting in healthier fish. Studies have shown that fish have a preference for jetties and reefs where cleaning stations are present.

CLEANING STATIONS: Sites where fish go to have parasites nibbled off by other fish or shrimp.

Snorkelling and dive sites

The upper reaches of the Gulfs tend to have few reefs, and those that are present are often relatively shallow and overlain by sand. The more exposed coasts of southern Yorke Peninsula and Fleurieu Peninsula, have numerous rocky shores and reefs.

Popular snorkelling sites include:

- Port Noarlunga – a great reef, close to Adelaide, with access from the end of the jetty.
- Second Valley – the bays on either side of the jetty offer protected snorkelling sites.
- Point Lowly – the water is cold in winter, but that's when the cuttlefish come to breed and they can be seen by snorkelling over the reefy shoreline.

Popular dive sites over reefs include:

- Aldinga Reef – dropping from 5 to 20 metres, the reef has lots of swim-throughs, different sorts of seaweeds and a great variety (and number) of fish.
- Seacliff Reef – a relict, submerged beach formation, lying in 12 metres of water. The reef is quite low, but its many holes and overhangs are home to lots of inquisitive fish.
- Haystacks – off the foot of Yorke Peninsula, is a very colourful dive with sponges, soft corals, an abundance of fish, huge bommies from 10 to 22 metres deep and ravines to swim through.
- Point Lowly – a shallow shore-dive (to 8 metres) offers the unique sight of cuttlefish breeding in winter.

Popular wreck dive sites include:

- Ex HMAS Hobart – sitting in 30 metres of water, and best dived at dodge tides when currents are reduced, the former navy destroyer offers a unique experience for advanced divers.
- Glenelg barge and the dredge (the South Australian) – lying at 20 metres depth, the barge and dredge are connected by a star-picket trail and both offer a variety of fish.
- Wardang Island (Port Victoria) wrecks – several wrecks are marked by buoys and lie in up to 8 metres of water. The Songvaar is the largest, and the Australian, McIntyre and Moorara are also popular.

More information about suitable dive sites is available from dive shops and dive clubs, and their websites.

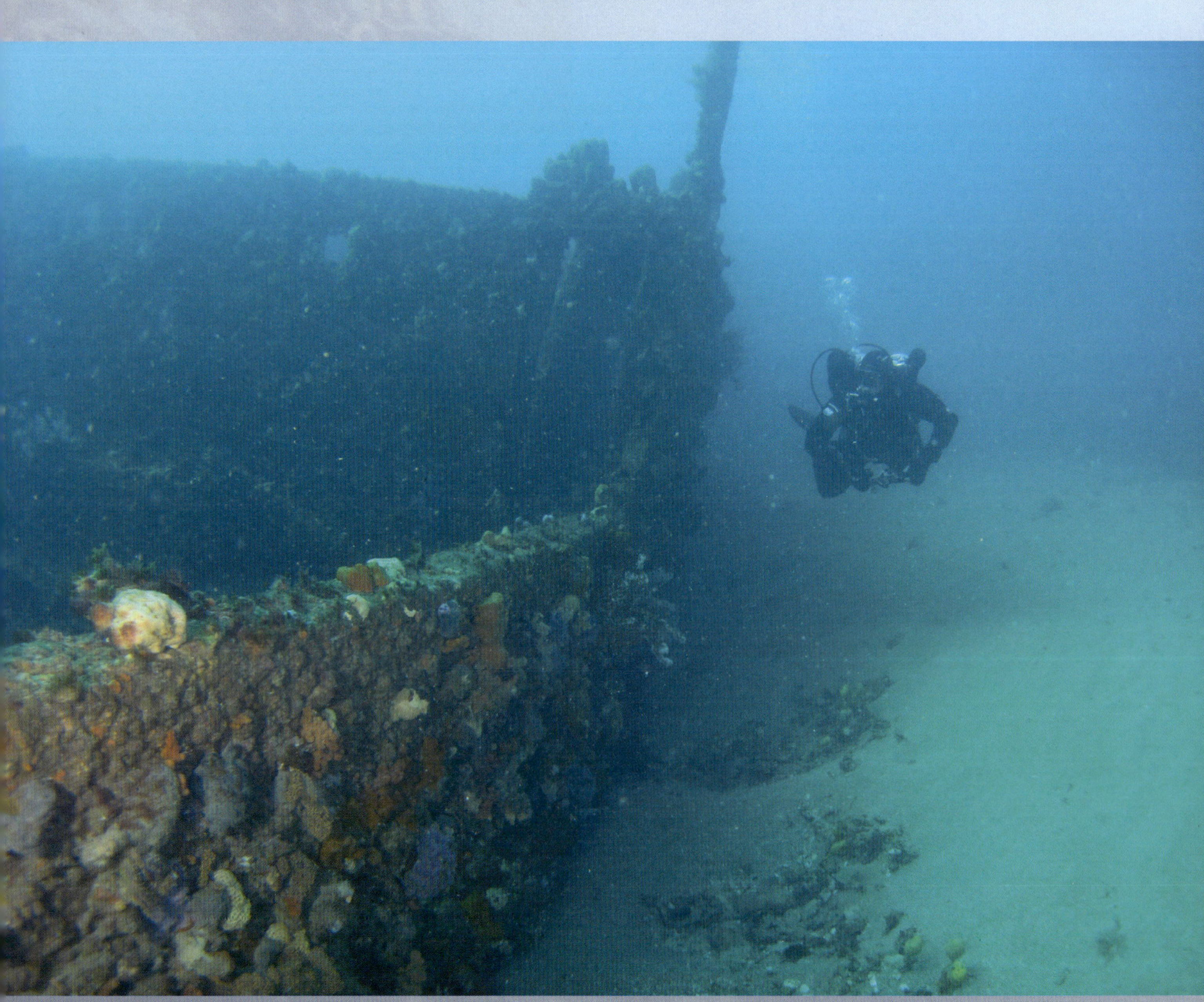

WHAT YOU'LL SEE

Common species of fish include:

- Sea Sweep
- Wrasse
- Herring Cale
- Moonlighters
- Leatherjacket
- Dusky Morwong
- Zebrafish
- Western Talma
- Magpie Perch
- Silver Drummer
- Bullseye
- Southern Blue Devil

REFERENCES

Connell SD & Gillanders BM. (2008) 'Marine Ecology'. Oxford University Press. South Melbourne. Victoria

– Chapter 14. Connell SD. 'Subtidal Temperate Rocky Habitats'

Shepherd SA, Bryars S, Kirkegaard IR, Harbison P, Jennings JT. (2008) 'Natural History of Gulf St Vincent'. Royal Society of South Australia Inc.

– Chapter 20. Turner DJ & Collings G. 'Subtidal macroalgal communities of Gulf St Vincent'

– Chapter 22. Shepherd SA & Baker JL 'Reef fishes of lower Gulf St Vincent'.

JETTIES

ECOLOGY OF JETTIES

Jetties provide a unique environment for marine creatures. They shade the water and lower the intensity of light, making conditions more like those of deep water. Jetty piles provide a hard vertical substrate for biota to grow on and a degree of protection for many species (including fish).

As an example, the old Rapid Bay jetty has one of the highest recordings of fish species for any site surveyed in the Gulf St Vincent. This is due to the shading effect from the jetty, the shelter provided by structures and debris under the jetty favouring deeper water fish, a wide range of micro-habitats providing niches for shallow water species, and a number of 'cleaner stations' that keep fish free of parasites.

Vertical surfaces, like jetty piles, tend to be covered by invertebrates, whereas level or sloping surfaces (as often associated with reefs) usually support light-dependent algae – so jetties are alive with sea-squirts, sponges and other colonial animals (which are discussed in later chapters).

INVERTEBRATES

A lot of marine invertebrates have free-floating eggs and larval stages that are carried by currents and tides as microscopic plankton. If they don't end up as food for fish or other invertebrates, their ultimate survival may depend upon them settling in a suitable site – one that is not already occupied for a start. To mature and reproduce, they need adequate food supplies and to be in the company of their kind (for sexual reproduction). Predation, competition for resources (such as food or space to grow) and seasonal conditions, can lead to fluctuations in the occurrence of many species.

BIOTA: All living organisms – plants and animals.

The animals attached to jetty piles are predominantly filter-feeders – they filter their food from water. Some thrive in slow currents (and may be vulnerable to damage in violent waters) while others need considerable water movement over them (and are armoured to withstand rough conditions). As an example, gorgonian fans like sites with moderate, bi-directional movement from surging waves.

Many of the organisms are colonial or modular – assemblages of small, repeated units, which are called zoids in ascidians and bryozoans, or polyps in corals. They generally reproduce both asexually (growing the colony by cloning) and sexually (through free-living larvae that aid dispersal and the colonisation of new sites). Chemical cues from other organisms are often a factor in whether larvae settle and attach or not – and chemicals are also used to discourage predation and as protection against infections.

SESSILE: Animals and plants that are fixed onto something and so are sedentary.

The jetties at Edithburgh and Rapid Bay have been extensively studied, revealing that at the small scale there is constant change in the species at any site, but at the larger (whole jetty) scale, the relative abundance of species is fairly constant. Edithburgh jetty is dominated by sponges (often growing as a colourful layer on scallops), while at Rapid Bay, hard corals (e.g. Culicia, see photograph below left) are more common.

Sessile animals compete for space through a variety of means. Some bryozoans nibble neighbours or settling larvae, some sponges produce toxic chemicals, and some colonial ascidians grow very quickly and smother their competitors. This diversity makes it hard for any one species to dominate. A fast grower may out-compete a slower species, but be retarded by chemical warfare from another – which may be vulnerable to nibbling by another species. In terms of their capacity for overgrowth, ascidians tend to outgrow sponges, which exceed bryozoans, which can dominate tubeworms – but rates of larval recruitment go in the opposite order. Good competitors may be poor recruiters.

Sea stars (also known as starfish) are major predators on some sessile species but there is little evidence of them having a significant impact on jetty biota. Some nudibranchs are also known to specialise in feeding on sponges but no grazing impacts have been studied.

Like reefs, jetties are not self-contained environments. They cannot be considered in isolation of their surrounds and features such as the supply of suspended food, damaging storms, and the arrival of schools of fish.

Snorkelling and dive sites

The State's history as an exporter of wool, grain and minerals, as well as the early coastal trade of ketches and fishing fleets, has produced some outstanding diving and snorkelling sites. Popular sites include jetties at:

- Rapid Bay – renowned for leafy sea dragons, if you're lucky enough to see one, but still a great dive even without them due to the variety and amount of fish life.
- Edithburgh – lots of colourful sponges, colonial ascidians and large sea-squirts, and an interesting collection of (often territorial) fish.
- Port Hughes –a highly rated dive spot, holding fish and a tremendous variety of invertebrates. Being shallow (about 5 metres), divers are able to spend an extended time underwater.

More information about suitable dive sites is available from dive shops and dive clubs, and their websites.

Sponges growing on Doughboy scallops (Edithburgh jetty)

What you'll see

Although small, many of the sessile organisms on jetties are spectacularly colourful and exquisite in formation. Jetty diving is best done slowly, taking time to explore the diverse creatures that reside (and visit) there.

You're likely to see:

- Sponge
- Sea Sweep
- Nudibranch
- Sea Squirt
- Dusky Morwong
- Western Talma
- Sea Stars
- Old Wives
- Yellowtail
- Coral
- Goatfish
- Wrasse

References

Shepherd SA, Bryars S, Kirkegaard IR, Harbison P, Jennings JT. (2008) 'Natural History of Gulf St Vincent'. Royal Society of South Australia Inc.

– Chapter 21. Butler A. 'Fauna of jetty piles, artificial reefs and biogenic surfaces'

– Chapter 22. Shepherd SA & Baker JL. 'Reef fishes of lower Gulf St Vincent'

SEAGRASS MEADOWS

ECOLOGY OF SEAGRASSES

Seagrasses are flowering plants, referred to as grasses because of their grass-like habit, although they are not true grasses. Their leaves are similar to blades of grass and they have an underground root (a rhizome) that stores energy and is involved in nutrient uptake and gaseous exchange – but they are not members of the Gramineae (grass) family.

The distribution of the various seagrasses is controlled by depth (light availability), substrate (e.g. a sandy bottom), wave exposure, nutrient supply, temperature, salinity and (for some) tidal inundation. The shallow, sheltered gulfs, with sandy and muddy floors, offer ideal growing conditions. Seagrass meadows typically extend to 15-20 metres depth, but may occur down to 40 metres. Posidonia grows to around 10 metres in upper Gulf St Vincent, 20 metres in the lower Gulf, and 30 metres in Investigator Straight.

There are very high levels of production and nutrient cycling in seagrass meadows but the plants themselves are not a valuable food resource. The gulf waters aren't home to big populations of dugong or other large browsers that can digest them. Seagrasses are more valuable for the habitat they provide to animals (which in turn serve as food for other animals), the service they provide in stabilising sands, and their food value when decomposing. They provide shelter and habitat for a rich web of species, including the juvenile stage of many types of fish.

The main seagrasses in the gulfs are: tapeweed (Posidonia), wireweed (Amphibolus), garweed (Zostera), paddleweed (Halophila) and Heterozostera. Tapeweed is most likely to form extensive meadows. It and wireweed are the longest lived and the ones most able to form a sheltering canopy. Posidonia traps sediments and organic detritus and, over several thousand years, builds up a fibre mat that may be several metres thick. Fibrous sheaths from their leaves can be rolled together by water motion and get washed onto beaches as 'posidonia balls' while their leaves form dense beds on some beaches, referred to as 'wrack'.

An extensive array of organisms, referred to as epiphytes, live attached to the leaves of seagrasses. They include algae, fungi and invertebrates, such as sponges, bryozoans, hydroids, worms and ascidians. Other invertebrates, such as crustaceans, molluscs and worms, live among the rhizomes.

The volume and variety of invertebrates in seagrass meadows provide rich pickings for fish and other organisms. Detritus (decaying organic material) washed from seagrass meadows also feeds invertebrates, and microbes, in adjacent reefs and mudflats. Although not significant in terms of grazing pressure, some organisms (including garfish, various leatherjackets, crabs, urchins and snails) consume seagrass directly. Leatherjackets tend to excrete seagrass intact – minus the invertebrate epiphytes that are their real food target.

RHIZOME: A root-like stem, usually horizontal and buried, which sends out roots below and shoots above.

The canopy of seagrass meadows also provides excellent protective cover for many species, including the juvenile stage of fish that may later migrate to deeper waters. Some species also rely on seagrass as a spawning area. King George whiting, garfish, southern calamary, razor-fish and blue swimmer crabs are often found in association with seagrass. Seahorses and pipefish are also commonly found among seagrass, with leafy sea dragons found on the margins of seagrass beds and reefs.

Being flowering plants, seagrasses can reproduce sexually, but they can also reproduce asexually by the cloning of rhizome segments. Amphibolus has separate male and female plants, but the other species have both sexes on the same plant.

SNORKELLING AND DIVE SITES

Seagrass beds can be found off many beaches in the Gulfs. Although they have been affected by urban pollution along the metropolitan coast, efforts are under way to reduce pollution loads and to restore these habitats. Divers can explore seagrass meadows in their own right or enjoy them as an adjunct to a dive on an adjacent reef, wreck or jetty. Examples include:

- Rapid Bay: The end of the jetty lies in seagrass and fish life observed there may include species that shelter near the jetty and make forays out over the seagrass as well as those typically found in seagrass.
- Wardang Island: Wrecks at the southern end of Wardang Island (the Australian and especially the McIntyre) lie on the edge of extensive seagrass beds that host a variety of well camouflaged fish and crabs.

WHAT YOU'LL SEE

- Razor-fish
- Goatfish
- Silverbelly
- Squid
- Stingrays
- Sponges
- Sea Dragons
- Sea Tulip
- Skates
- Yellowtail
- Blue Swimmer Crab
- Sea Cucumber

REFERENCES

Connell SD & Gillanders BM. (2008) 'Marine Ecology'. Oxford University Press. South Melbourne. Victoria

– Chapter 17. Gillanders BM. 'Seagrass'

Shepherd SA, Bryars S, Kirkegaard IR, Harbison P, Jennings JT. (2008) 'Natural History of Gulf St Vincent'. Royal Society of South Australia Inc.

– Chapter 11. Bryars S, Wear R & Collings G. Seagrasses of Gulf St Vincent and Investigator Straight'

– Chapter 12. Jones GK, Connolly RM & Bloomfield AL. 'Ecology of fish in seagrass'

ROCKY SHORELINES

ECOLOGY OF SHORELINES

Rocky shorelines and beaches are where the land meets the sea, and where most people get their first glimpse of the underwater world. They are home to a wide diversity of specially adapted small creatures, which can be found with some patient fossicking.

This is a tough environment in which to live. The shore may be underwater sometimes or exposed to the air at others. It can be buffeted by strong waves and winds, salinity may be very high as salt water evaporates or quite low as rain forms pools, and it may be exposed to high levels of ultra violet (UV) light and high temperatures. It is, however, highly productive thanks to the good light levels and supplies of nutrients (coming from the sea and the land).

Many of the invertebrates on rocky shores are seasonal breeders and may release high numbers of larvae or eggs in a single event, creating rich plankton that becomes food for other organisms. There is usually high mortality and heavy predation for larvae, but sometimes there may also be a sudden population boom after a particularly successful breeding event.

Port Moorowie

Barker Rocks

Port Noarlunga

Shoreline zones

The height of the shore relative to the average sea level can determine which plants and animals survive.

The highest parts of shorelines are mainly dry and highly saline due to sea spray – although occasionally they are inundated by the highest of tides or flushed by heavy rainfall. Lichens may cover the rocks and molluscs may graze upon them – relatively free from predators due to the harsh conditions.

Lower down, between the average high and low water levels, there are more diverse microhabitats available and increased numbers of species and creatures taking advantage of them. Algae such as the green sea lettuce (ulva) may persist in pools and invertebrate animals may include tubeworms and barnacles that feed by filtering plankton from the water with fine meshed appendages. Molluscs such as chitons and snails, crustaceans including crabs and shrimp, and sea stars and anemones can often be found.

Rocks may have hardy species growing attached to their upper surface, while more delicate sponges or sea squirts (Ascidians) grow attached to their underside, protected from the desiccating sun, wind and UV light during low tide. If rocks are found disturbed and upturned it is best to replace them right-side up to protect the organisms on their shaded under-side.

Intertidal rock pools hold water at low tide and can maintain small populations of fish. Dropping a morsel like a scrap of fishing bait into a pool may draw out some shrimp from hiding, then a small fish that will see the shrimp off, and perhaps a crab (the most common predator) which will scare off the fish before claiming the treat as its own.

The intertidal zone may also be home to the small, but very venomous, blue-ringed octopus. Take care of the environment and yourself, when investigating rock pools.

Second Valley

Slippery microbial films may cover the rocks closer to low water and be grazed by limpets and snails. At the lowest of tides, areas of seaweed that are normally underwater may occasionally be exposed to the air, wind and sunshine. The type of rock (e.g. very hard, smooth granite or soft, coarse limestone) will influence the type of seaweeds to attach and grow.

Below the low water mark, seaweeds dominate the landscape. The golden flat fronds of kelp are a common sight, forming a canopy that provides habitat for a bewildering array of sea life. In South Australia the kelp tends to be shorter than in eastern Australia, where it may be up to a metre long. It often has an understory of hard pink, encrusting coralline algae – looking like the rocks have been painted. Storms, or pollution, may damage the kelp beds, opening the way for low, sediment trapping algae to take their place until the canopy can regrow and shade them out.

Snorkelling and dive sites

Accessible rocky shores and occasionally exposed reef tops can be explored at:

- Hallet Cove
- Port Noarlunga
- Point Lowly
- Lady Bay
- Second Valley
- Myponga Beach
- Marion Bay
- Barker Rocks / Parsons Beach
- Point Gilbert / Port Moorowie

What you'll see

- Periwinkle
- Limpet
- Tubeworm
- Barnacle
- Chiton
- Crab
- Sea Star
- Shrimp
- Grubfish
- Anemone
- Sea Lettuce
- Neptune's Necklace
- Robust Zoanthid

References

Connell SD & Gillanders BM. (2008) 'Marine Ecology'. Oxford University Press. South Melbourne. Victoria

– Chapter 15. Underwood AJ & Chapman MG. 'Intertidal Temperate Rocky Shores'

Shepherd SA, Bryars S, Kirkegaard IR, Harbison P, Jennings JT. (2008) 'Natural History of Gulf St Vincent'. Royal Society of South Australia Inc.

– Chapter 10. Benkendorff K, Fairweather P, Dittman S. 'Intertidal shores: life on the edge'

Kingdoms of the sea

Evolutionary overview

Life on earth began in water and evolved in the sea, before venturing onto land. For billions of years, the Earth's oceans spawned new life forms and many of their descendants survive today. Underwater it is possible to see organisms that are living relatives of some of the first life forms on earth. Many can only be found in the sea – there is simply nothing like them on land. Their fascinating shapes and colours are only matched by their biology and the story of their evolution.

To help describe the diversity of organisms, and relationships between them, scientists classify them first into kingdoms, which are split into phyla (for animals) or divisions (for plants); and then into classes, orders, families, genus and finally species. Divers and snorkellers will regularly see representatives of the algae (seaweeds), vascular plant (seagrasses) and animal kingdoms, including some of the most ancient life forms still alive.

Useful web references, including phyla not covered in this book, include:

Australian Museum:	australianmuseum.net.au/animals
Marine Education Society of Australia:	www.mesa.edu.au/default.asp
Reef Watch:	www.reefwatch.asn.au
Woodbridge Marine Discovery Centre:	www.woodbridge.tased.edu.au/MDC
Marine Life Society SA – marine index:	www.mlssa.asn.au/marine.html
Marine Species Identification Portal:	species-identification.org/index.php

Scientific terms for classifying organisms

Kingdom

Phylum (animals) or Division (plants)

Class

Order

Family

Genus

Species

Scientific names are given as Genus species, e.g. *Phycodurus eques* – the leafy sea dragon, South Australia's marine emblem. Scientific names are derived from Latin and ancient Greek.

The most commonly viewed kingdoms of the sea, and their main phyla and divisions, are introduced on the next page, with some regard to their evolutionary order, as indicated in the diagram above.

Many of the earliest animal life forms are invertebrates that are still only found in aquatic (primarily marine) environments, such as Porifera and Cnidaria. Echinoderms and Urochordata are also exclusively aquatic, but other invertebrates (Mollusca, Annelida and Anthropoda) are marine and terrestrial. Vertebrates such as mammals and birds (Aves) developed on land, but some have migrated back to the oceans.

TAXONOMY

The naming and classification of organisms is referred to as taxonomy, and it is a dynamic science. As new information and evidence comes to hand, e.g. from DNA analysis, scientists revise their thinking about the relationships between species and may rename or re-group species, families or even whole phyla.

There may be significant debate before new classifications are accepted. We have tried to dodge those debates in this book and have generally opted for traditional or widely used classifications for consistency with commonly available texts.

Useful taxonomic web references include:

– World Register of Marine Species (WoRMS): www.marinespecies.org/index.php

– Tree of Life: tolweb.org

– University of Edinburgh Natural History Collections: www.nhc.ed.ac.uk

Major plant and animal groups

Algae (seaweeds)

- Chlorophyta: green algae
- Phaeophyta: brown algae
- Rhodophyta: red algae.

Angiosperms (flowering plants, including sea-grasses)

Porifera (sponges)

Cnidaria

- Hydrozoa: hydra
- Scyphozoa: jellyfish
- Anthozoa: including the Orders Actinaria (anemones) and Zoantharia (corals).

Bryozoa (lace corals and moss animals)

Mollusca (shelled animals)

- Polyplacophora: chitons.
- Gastropoda: snails, slugs and limpets
- Bivalvia: scallops and mussels
- Cephalopoda: cuttlefish, squid, octopuses and nautilus

Annelida (worms)

- Polychaeta

Arthropoda:
Crustaceans – the main marine phylum, including the classes of:

- Cirripedia: barnacles
- Malacostraca: lice and shrimp, prawns, crayfish and crabs from the Order Decapoda.

Echinodermata

- Crinoidea: feather stars
- Asteroidea: sea stars
- Ophiurodea: brittle stars
- Echinodea: sea urchins
- Holothuroidea: sea cucumbers

Chordata:
Urochordata – tunicates*

- Ascidiacea: sea squirts

Vertebrata, including:

- Chondrichthyes: sharks and rays, which have a cartilaginous skeleton
- Osteichthyes: bony fish and sea horses
- Aves: birds
- Mammalia: dolphins, seals and humans.

** Although there are rudimentary similarities between tunicates and other chordates, there are significant differences and some modern taxonomies describe the tunicates as a separate phylum: Tunicata.*

Kingdoms of the Sea

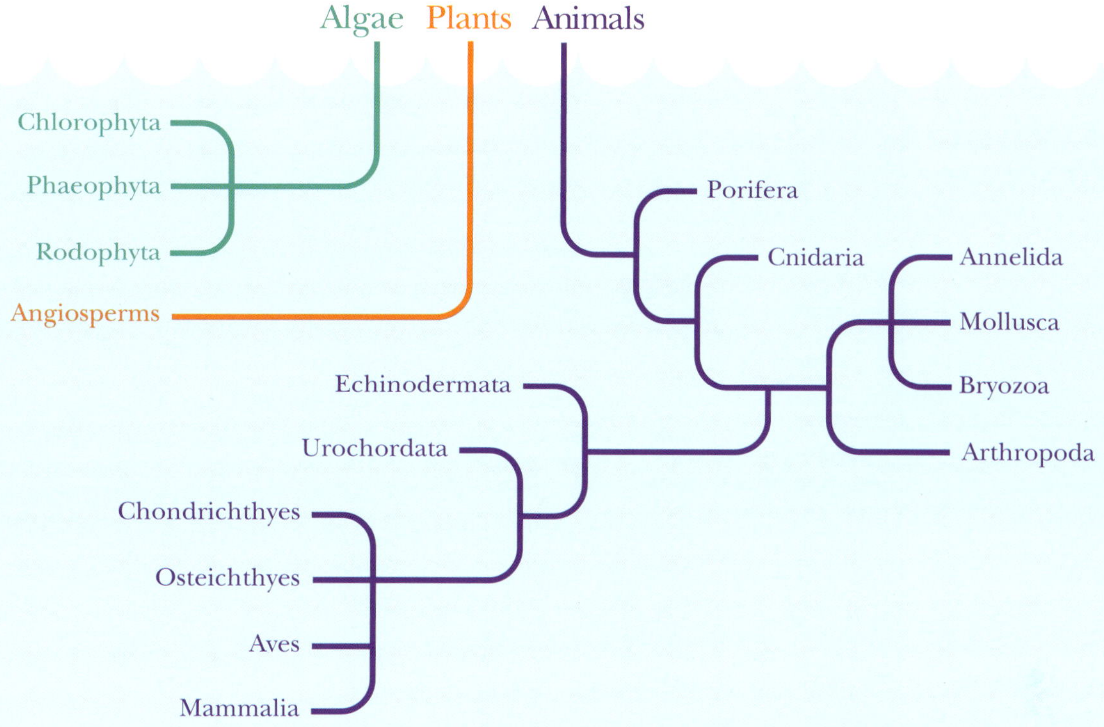

Algae Plants Animals

Chlorophyta
Phaeophyta
Rodophyta
Angiosperms
Porifera
Cnidaria
Annelida
Mollusca
Bryozoa
Arthropoda
Echinodermata
Urochordata
Chondrichthyes
Osteichthyes
Aves
Mammalia

These creatures are discussed in more detail in the following sections and some commonly seen species are introduced with pictures.

On the land, we are used to seeing mobile creatures that seek out plants or other animals as food. The same applies in the sea. However, in the marine environment there are also a lot of stationary (sessile) creatures who dine on whatever the currents bring to them. They may filter microscopic bacteria or plankton from the water or actively catch and consume it. The apparatus to permit these activities – siphons to channel water, fine nets to waft through the water, and 'tentacles' to capture prey – may be seen in a variety of different animal groups. Their genealogy may be quite different, but their mode of feeding and appearance may be similar. Some of the most ancient life forms (sponges) may be hard to tell from more recent ones, such as Ascidians.

Telling one creature from another, and correctly identifying both, is not always easy underwater. The following pictures and notes continue the discussion of the evolution of marine creatures and aim to make it easier to identify them.

ALGAE

SEAWEEDS

PHOTOSYNTHESIS: The creation of organic matter and oxygen from carbon dioxide, water and the sun's energy, by cells containing the green pigment, chlorophyll.

Single-celled algae were among the earliest of life forms on Earth and played a major role in changing the atmosphere – converting carbon dioxide to oxygen through photosynthesis. They are still incredibly important, through their roles in generating oxygen and as food for planktonivores and herbivores. They are the foundation of many marine food-webs and crucial to the earth's atmosphere.

An evolutionary progression saw single-celled algae assemble into multi-cellular organisms, followed by differentiation between cells to form more structured seaweeds. Commonly described divisions within the algal Kingdom include:

- Chlorophyta (green algae)
- Phaeophyta (brown algae)
- Rhodophyta (red algae).

More information on algae is available at the Seaweed Site, www.seaweed.ie/index.html

1. Sea lettuce
Ulva sp.

2. Neptune's Necklace
Hormosira banksii
'A bead a day' was once recommended as a health supplement due its high iodine content.

3. Warty Seaweed
Scaberia agardhii
Several brown seaweeds have numerous rounded, ball-like structures that act as floats and may contain reproductive structures.

4. Brown Seaweed
Cystophora botryocystis

5. Common Kelp
Ecklonia radiata

6. Red Algae
Plocamium sp. (suspected)

SEAWEEDS YOU MIGHT SEE

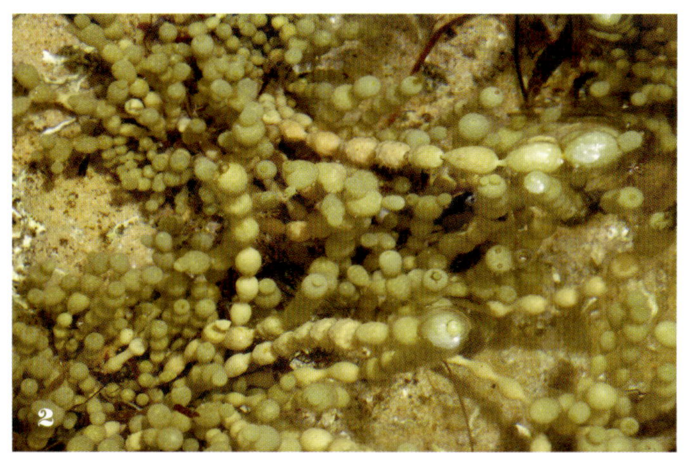
2

3

4

5

6

ANGIOSPERMS

FLOWERING PLANTS – SEAGRASSES

Early plant forms, such as mosses, liverworts and ferns, grew on land, but in moist conditions – and sometimes still relied on the presence of water for reproduction to occur. The evolution of 'modern' plants was marked by the development of seeds in Gymnosperms (cone-bearing plants) and Angiosperms (flowering plants). Flowering plants evolved on land and 'seagrasses' are a rare example of flowering plants that have adapted to the marine environment.

1

Seagrasses you might see

1. Paddleweed *Halophila australis*

2. Tapeweed *Posidonia australis*
With a school of King George whiting (Sillaginodes punctatus)

3. Garweed *Zostera sp*

4. Wireweed *Amphibolus antarctica*

PORIFERA

SPONGES

Sponges are some of the most ancient of animal life forms, with fossil records dating back more than 570 million years. They are simple creatures with an outer and inner layer, but no internal organs. The cells of the inner layer have 'flagella' that waft water through the sponge and trap bacteria and other suspended food from the water. Water is taken in through many small pores and expelled through a small number of large openings in the sponge.

Sponges come in an array of shapes, sizes and colours, ranging through encrusting mats, balls, fans, tubes, cups and tree-like forms. They are supported by combinations of calcium or silica-based structures (spicules) and fibres made of protein (spongin) which form a type of skeleton that may be found washed up on beaches once the sponge has died. Large, erect forms have more skeletal structures and are slow growing, but long lived. Encrusting species grow rapidly and can quickly recolonise areas following a disturbance. Many sponges produce chemicals that are toxic or distasteful to predators, although some species have evolved to counteract those toxins.

Sponges are often hermaphroditic and reproduce by spawning. The resultant free-swimming larvae then settle and grow into new sponges. Some can also reproduce asexually by budding.

FLAGELLA: Thin threads that protrude from cells and can move in waves or a whip-like fashion.

HERMAPHRODITES: Animals with male and female reproductive organs in the same individual. Self fertilisation is avoided by measures such as releasing sperm at different times to the eggs.

SPAWNING: The release of eggs and sperm into water where fertilisation occurs.

1. ***Suberites sp.***
Often found partly covered by sand.

2. Unidentified Sponge

3. & 4. ***Holopsamma sp.***
Encrusting sponges may die back during summer and regrow in cooler months. Some species may be grazed by nudibranchs.

5. Sycon
Sycon sp.

6. ***Aplysilla rosea***

7. Cup sponge ***Cymbastela sp.***

8. Ball sponge ***Ancorina sp.***

Sponges you might see

CNIDARIA

POLYPS AND JELLYFISH

Cnidaria are the 'stingers' of the sea – though only a relatively few in South Australia are harmful to humans. They have a simple structure including a central 'mouth' leading into a body cavity where food is digested. Their inner and outer layers are separated by a jelly-like substance of variable thickness, being thickest in jellyfish. All Cnidaria have stinging tentacles around the central 'mouth' (through which waste is also expelled). The 'sting' comes from specialist cells (nematocysts) that fire barbed harpoon-like threads into victims, and inject toxins or adhesives at the same time. The nematocysts are for protection and to capture prey, and are unique to cnidarians.

Many Cnidaria have both a fixed polyp (or sack-like) form (e.g. anemones) and a free swimming medusa (bell-shaped) form (e.g. jellyfish) at different stages in their life, although one form may be dominant. They can breed by cloning (asexual budding) or by spawning.

CHITIN: A nitrogen-based organic compound that builds strong coverings and structures in animals.

Cnidaria are carnivorous, but some also host algae or other photosynthetic organisms within their bodies, from which they can gain energy. They can live individually or in colonies and many develop skeleton like features from chitin or calcium carbonate (e.g. reef-forming corals).

The main classes of Cnidaria are:

- Hydrozoa: hydra
- Scyphozoa: jellyfish
- Anthozoa: including the Orders Actinaria (anemones) and Zoantharia (corals).

Corals are usually thought of as tropical species but some are at home in cooler waters, especially some of the soft corals (which do not form hard calcareous outer casings). There are also some species in South Australia that are thought to be relicts from times when local seas were warmer.

CNIDARIANS YOU MIGHT SEE

1. Jellyfish *Pseudorhiza haeckeli*
Capable of giving a mild sting, this jellyfish has nematocysts on the bell as well as its tentacles. It is sometimes found washed onto beaches.

2. Anemone *Actinia tenebrosa*
Anemones may appear to be cemented in place but they can release their hold and move.

3. Hydroid fern *Pennaria disticha*
The wispy white fronds look like a fern but are actually an animal.

4

4. Robust Zoanthid
Zoanthus robustus
Their green colour comes from photosynthetic algae in their bodies and their tentacles are used to catch plankton, especially at night.

5. Carijoa Soft Coral
Carijoa
Thought to be a tropical relict, covered here by an encrusting orange sponge.

6. Soft Coral
Erythropodium hicksoni

7. Solitary Stony Coral
Scolymia australis

8. Gorgonian fan
Mopsella zimmeri

BRYOZOA

LACE CORALS AND MOSS ANIMALS

Bryozoans are colonial animals – individuals living as a group – which can look like anything from moss to fine, tiny lace-work. The individuals are very small (a few millimetres at most) but each secretes a case around itself and they collectively form a colony. They grow on rocks and jetty pilings, and on other animals and plants.

Each tiny individual has a set of sticky tentacles (termed a lopophore) that are used to lasso food. However, because they have the ability to share food between themselves (via connecting tissue) not all animals have to be involved in harvesting. Some animals in the colony may specialise to develop beak-like structures to ward off enemies or long bristles to brush away detritus. Some even develop into brood chambers. Reproduction is mainly by asexual budding, but larvae are also produced. Like sponges, some bryozoans protect themselves with unpalatable chemicals.

DETRITUS: Decomposing organic matter from plants and animals.

BRYOZOANS YOU MIGHT SEE

1. **Lace Coral** *Triphyllozoon sinuatum*
2. **Tubular Bryozoan** *Celleporaria sp.*

2

MOLLUSCA

SNAILS, SLUGS, SCALLOPS AND SQUID

The Mollusca phylum is one of the most diverse and extensive groups of animals, with representatives on land and in water. They are soft bodied but often covered by a hard shell. They come in a variety of forms, but usually have a 'head' and 'foot' region and some have sensory 'feelers' and eyes. They have internal organs such as a gut, kidneys, gonads, a heart, gills and a nervous system. Many graze on plants and encrusting animals (such as sponges, bryozoans and ascidians) and have a 'radula' – a rasp-like tongue used in feeding.

They include:

- Polyplacophora: chitons.
- Gastropoda: snails, slugs and limpets
- Bivalvia: scallops and mussels
- Scaphopoda: tusk shells
- Cephalopoda: cuttlefish, squid, octopuses and nautilus

Chitons have eight overlapping plates on their back, and a fleshy foot that is lined with a row of gills. Some are able to roll up into a ball as a form of protection. There are males and females, and fertilisation to form larvae occurs in the water.

Gastropods range from grazers and scavengers, to filter feeders, to specialised predators able to drill holes into the shells of other gastropods or crustaceans. The sexes are separate in many species, but some are hermaphrodites and some species change sex during their lifetime.

Nudibranchs (or sea slugs) are an unusual and colourful subset of gastropods that lack a shell in their adult form. Most have retractable tentacles and some have ornate exposed gills. Grazing on encrusting invertebrates, many nudibranchs develop special means of dealing with the toxins they ingest, incorporating them in their own bodies and becoming toxic to creatures that may otherwise prey upon them. Their colourful form

Squid eggs washed onto the beach.

1. Squid (Southern Calamary)
Sepioteuthis australis

advertises their toxicity and warns potential predators away. A sub-order of nudibranchs (the Aeolidina) graze on cnidarians and concentrate their stinging nematocysts into external sacs where they remain active and protect the nudibranch.

Bivalves (e.g. scallops) typically have two shells and are filter feeders. They do not have a radula, but may have numerous sensory tentacles and simple eyes to detect danger. They may grow anchored to something or, as in scallops, be able to 'swim' away from danger.

Cephalopods have advanced nervous systems, including very well-developed eyes, and their form is quite different to other molluscs.

- The 'shell' has become an internal 'backbone' in cuttlefish and squid, and is absent in octopods.
- The 'foot' of other molluscs has developed into a ring of tentacles around the mouth, and they usually have rows of suckers on each.
- Their mouth includes a parrot-like beak as well as a horny radula.

Many cephalopods can change their colour, and even shape, to mimic their surroundings or for communication. There are usually distinct male and female forms. Following mating, the females lay white egg sacks that are secured under ledges and are sometimes found washed onto beaches after a storm.

2. Razorfish
Pinna bicolour
The razor sharp edges must be handled with care, these shells are often coated by plants and animals.

MOLLUSCS YOU MIGHT SEE

3. Chiton
Rhyssoplax calliozona

4. Nudibranch
Ceratosoma brevicaudatum

5. Abalone
Haliotis laevigata
A row of holes near the edge of the shell are for respiratory (breathing) purposes.

6. Giant Cuttlefish
Sepia apama

Cuttlefish are able to change shape and colour, including flashing with strobes of colour.

7. Doughboy Scallop
Mimachlamys asperrima
Often covered in colourful sponges, the scallop uses its blue eyes to detect threatening movements.

Giant cuttlefish

Giant cuttlefish (Sepia apama) can be found almost anywhere in the Gulfs, but the area west of Point Lowly (near Whyalla) is renowned internationally for the mass breeding every winter. Thousands of giant cuttlefish provide an intriguing display for divers and snorkellers. A local fishing ban now provides protection in the immediate vicinity of this amazing event.

Cuttlefish have some incredible physical characteristics matched only by their social antics. They are masters of camouflage and can change shape as well as colour to match their surroundings and for communication to attract mates or warn off competitors. They are typically smooth skinned, but can draw their skin up into an array of knobs and bumps to mimic seaweed, or turn themselves into fearful-looking 'monsters' with flashes of red to highlight their hostile appearance.

Using 'jet propulsion' they can shoot backward as they squirt out a stream of water (and maybe a screen of black ink – 'sepia'), but they can also move forward using the wave-like action of their fringing mantle. Hunting tentacles can be shot forward at amazing speed.

They live for 1-2 years and will only breed once, after which they die. Mating is a competitive affair. During 'courtship' large males grapple for the right to a female – and a smaller male may take advantage of the distraction to pay a quick visit to the object of their desire. After mating, males guard the female, who attaches her eggs – a mass of individual white globules from which tiny adults later emerge – to the roof of caves and under ledges.

During mating season the cuttlefish are almost oblivious to divers and the social displays are easily observed and interpreted, without interruption.

ANNELIDA

WORMS

Annelida are segmented worms and most species are now classed as Polychaetes. They have a head, a 'tail-end' (referred to as the pygidium), and are made up of repeated segments (each with breathing and excretory organs, plus 'parapoda' – muscular projections that usually bear hair like structures referred to as setae, which are used in movement).

Some species are mobile scavengers, and some are predators with extendable jaws and teeth like the creatures in 'Alien', but many species live sedentary lives inside protective casings which they secrete. The tube-dwelling worms have specialised retractable gills that protrude from the tube and are used for respiration or filter feeding, e.g. fan worms. Tube worms are filter feeders which have evolved to survive in the intertidal zone and have a covering to seal off their calcareous tubes when exposed to the air.

Although some are hermaphroditic, Polychaete worms usually have separate sexes and reproduce by spawning. Some species can reproduce asexually by budding off segments.

WORMS YOU MIGHT SEE

1. Inter-tidal Tube Worm
Galeolaria caespitosa

2. Australian Fan Worm
Sabellastarte australiensis
Fan worms are very sensitive and rapidly retract the feathery feeding crown if disturbed. An introduced pest, the European fan worm, is similar but has an elevated spiral crown and occurs in groups.

1

2

ARTHROPODA

CRABS AND CRAYFISH

Arthropods are found on land and in the water, and include insects, spiders and crabs. They have a stiff outer covering (an exoskeleton), segmented bodies with a head and abdomen, and segmented legs. Other body parts are also segmented and may be modified into antennae, mouthparts, and reproductive organs. Their compound eyes can be very advanced structures. The phylum has been divided into three sub-phylum: Chelicerata (spiders, scorpions and mites), Uniramia (insects, centipedes and millipedes), and Crustacea (crabs, crayfish and shrimp).

Crustaceans are mainly marine and include the classes of:

- Cirripedia: barnacles
- Malacostraca: lice and shrimp, prawns, crayfish and crabs from the Order Decapoda.

Crustaceans usually have two pairs of antennae, although they are not always obvious. Most species have a planktonic larval form and some species are very small, remaining as plankton all their life. Crustaceans grow by shedding their exoskeleton and forming a new, slightly larger one.

Barnacles are heavily armoured filter feeders in which the legs have become feathery appendages that waft through the water, collecting food.

The **Decapoda** ('ten footed') have five pairs of limbs used in movement, usually claws and legs, although in some species several 'legs' may be modified for special purposes. In most decapods, the female carries her eggs held by swimmerets under the abdomen. The Order includes some of the most sought-after foods from the ocean.

1. Decorator Crab
Naxia aurita
A 'decorator crab' attaches seaweed to itself as camouflage

2. Shrimp
Palaemon serenus

ECHINODERMS YOU MIGHT SEE

4

5

6

7

8

9

1. Eleven-armed Sea Star
Coscinasterias muricata

2.
Echinaster arcystatus

3. Sea Cucumber
Stichopus ludwigi

4. Vermillion Biscuit Star
Pentagonaster dubeni

5.
Fromia polypora

6.
Nectria ocellata

7. Sea Urchin
Centrostephanus tenuispinus

8.
Petricia vernicina

9. Biscuit Sea Star
Tosia australis
Pictured on a sponge covered with white strands – the castings around Polychaete worms – which the Biscuit Sea Star is probably eating.

CHORDATA

Chordates have unique embryonic structures that may disappear with maturity: a dorsal nerve chord, a notochord (a rod-like structure near the nerve chord), and gills (which may develop into filter feeding devices, gills, or parts of the inner ear). They include:

Urochordata (or, in some taxonomies, the separate phylum Tunicata), including Ascidians, or sea squirts

Vertebrata: vertebrates (animals with backbones) including fish, birds and mammals.

UROCHORDATA

TUNICATES, SEA SQUIRTS

CELLULOSE: A major part of the cell walls of plants and many algae.

The tunicates are sack-like creatures that are usually attached to firm substrate and may be solitary or in tight colonies. They often have a leathery covering (the 'tunic'), two external openings (an inhalant and exhalent siphon), and an internal filter feeding structure. The 'tunic' is made of tunicin – a unique cellulose-like substance. Tunicates are hermaphroditic and have an unusual heart which pumps blood first in one direction then, after a while, the other.

Ascidians (class Ascidiacea) are the most commonly seen tunicates. They include solitary sea squirts, smaller colonial forms and solitary sea tulips – which resemble lumpy sea squirts on a long stem. Many ascidians produce distasteful or toxic chemicals and may be brightly coloured to warn predators away. Sea tulips may be covered in encrusting sponges – providing the sponge somewhere to live in favourable currents and giving the ascidian protection due to the sponge's chemical deterrents.

Solitary sea squirts are very muscular and able to squirt water out of their siphons – hence the generic name. Solitary sea squirts usually reproduce sexually by spawning, but colonial ascidians can also reproduce asexually by budding. In colonial forms, the individuals may be 'cemented' together and have interconnecting tissues. They may have a larger common exhalent opening – and with numerous small inhalant and a few large exhalent openings can resemble sponges.

ASCIDIANS YOU MIGHT SEE

1. Compound Ascidian
Leptoclinides sp.

2.
Didemnum delectum

3. Brain Ascidian
Sycozoa cerebriformis

4. Ball Ascidian
Polycitor giganteus

5. Colonial Sea Squirt
Clavelina moluccensis

6. Solitary Sea Squirt
Phallusia obesa

7. Sea Tulip
Pyura spinifera

8.
Botrylloides sp. (suspected)

5

VERTEBRATA

SHARKS, FISH, BIRDS & SEALS

The vertebrates include:

- Chondrichthyes: sharks and rays, which have a cartilaginous skeleton
- Osteichthyes: bony fish and sea horses
- Aves: birds
- Mammalia: dolphins, seals and humans.

Chondrichthyes have a skeleton made of cartilage, and have from five to seven gill slits. They include sharks, rays and skates and have ancestors dating back 450 million years. Their eyesight is often not strong but they have special sensory organs that respond to minute electrical currents and an advanced sense of 'smell'. Skates and rays are both flattened for life on the ocean floor, but skates do not have venom in their dorsal spines.

All sharks, rays and skates reproduce sexually through internal fertilisation. Sharks may lay eggs in distinctive egg cases (more commonly those that are bottom dwellers) or give birth to live young (more commonly mid-water species). Young grey nurse sharks are known to eat siblings while still in the womb.

DORSAL FIN: On the back. The dorsal fin runs along the back of a fish.

VENTRAL FIN: On the front or belly. Also referred to as the pelvic fin.

PECTORAL FIN: Pectoral fins lie behind the gills.

CAUDAL FIN: On the posterior. The tail fin.

1. Southern Banded Wobbegong
Orectolobus halei
Wobbegong are often temporary residents in an area before moving on. They are usually nocturnal, hunting at night and resting during the day.

SHARKS AND RAYS YOU MIGHT SEE

2. Port Jackson Shark

Heterodontus portusjacksoni

Port Jackson sharks return to breed in the same location and lay eggs in spiral egg cases (which may be found washed up on beaches). They have a venomous barb in front of the dorsal fin, and feed mainly on molluscs.

3. Banded stingaree

Urolophus cruciatus

Though relatively small (50 cm), stingarees have one or two venomous spines on their tail.

4. Smooth stingray

Dasyatis brevicaudata

Rays have no dorsal fins but a long thin tail with one or two venomous barbs. The ray pictured below appears to have lost the end of its barbed tail.

FISH YOU MIGHT SEE

Osteichthyes (fish and sea horses) have a bony (calcified) skeleton and their body is usually covered in scales. They have a keen sense of smell (like sharks and rays) but also have very good eyesight. They rely on a unique organ – the swim bladder – for buoyancy. Fish expand or contract the bladder, in the same way that a scuba diver uses a buoyancy vest, to avoid sinking or floating to the surface. Fish can waft water over their gills ('breathing') without having to swim.

Most fish reproduce sexually by spawning and some species can go through a number of larval forms before maturing.

Some fish are very inquisitive and will circle and perhaps even follow divers and snorkellers. Silver drummer, cowfish, young sweep and leather jackets are classically curious. Some other species are very territorial and will swim up to divers and 'get in their face' – regardless of the difference in size, e.g. Southern blue devils.

1. Western Talma
Chelmonops curiosus
Related to tropical butterfly fish, Talma feed on small invertebrates.

1

2

3

4

5

2. Old Wives
Enoplosus armatus
The dorsal spine of Old Wives is venomous. They separate into pairs prior to spawning.

3. Zebrafish
Girella zebra
Zebrafish nibble on seaweeds and are usually found in schools.

4. Moonlighter
Tilodon sexfaciatum
Moonlighters feed on benthic invertebrates.

5. Magpie Perch
Cheilodactylus nigripes
Magpie Perch are territorial. They filter small invertebrates from mouthfuls of sand. They, and other fishes, can sometimes be seen resting in the shelter of a sponge.

6. Dusky Morwong (Strongfish)
Dactylophora nigricans

7. Longsnout Boarfish
Pentaceropsis recurvirostris
These inquisitive fish can be found in very deep water, up to 260 metres, and have venomous spines near their fins.

8. Rock Flathead
Platycephalus laevigatus
Flathead are often seen lying in ambush, ready to pounce on passing prey.

9

9. Yellowhead Hulafish

Trachinops norlungae

Hulafish are found in large schools in sheltered sites, often in caves or under ledges.

10. Bullseyes

Pempheris klunzingeri

Bullseyes are nocturnal planktonivores, usually found sheltering in caves during the day.

11. Silver Drummer

Kyphosus sydneyanus

Silver drummer are inquisitive and schools will often follow or circle divers and snorkellers. They are herbivorous.

10

11

12

13

14

15

16

17

18

12. Southern Blue Devil
Paraplesiops meleagris
Blue devils can be defensive of their territory and try to ward divers off, before retreating to their home base.

13. Scalyfin
Parma victoriae
Scalyfin are gardeners; they remove unwanted seaweeds to maintain the ones they prefer as feed. They are territorial and the male also guards their eggs. Juveniles are more colourful than adults.

14. Sea Sweep
Scorpis aequipinnis
An inquisitive fish, sweep are usually found in small schools around reefs and some jetties.

15. Horseshoe Leatherjacket
Meuschenia hippocrepis
Leatherjackets do not have scales and can swim backwards using their fins.

16. Bluethroat Wrasse
Notolabrus tetricus
Bluethroated wrasse live in 'harems' of 1-2 males and 10-15 females. A female will change into a male if one dies.

17. Globefish
Diodon nicthemerus
Globefish are nocturnal, feeding on benthic invertebrates at night and sheltering during the day.

18. Goatfish
Upeneichthys vlamingii
Goatfish use the sensory barbels on their chin to detect food in sand.

19

20

19. Herring Cale

Olisthops cyanomelas

The herbivorous herring cale have a beautiful sinuous swimming motion. They keep their distance but will do 'swim-bys' to keep watch on divers.

20. Western Blue Groper

Achoerodus gouldii

A large, docile and inquisitive fish, the blue groper is a favourite with divers.

21. Shaw's Cowfish (female)
Arcana aurita
Cowfish are like inquisitive busybodies. They will swim toward and circle divers with their tiny fins a-flutter, checking out the newcomers to their territory.

22. Common Gurnard Perch
Neosebastes scorpaenoides
The dorsal spines of the Gurnard Perch are venomous and the fish are often encountered lying concealed on the ocean floor.

23. Queen Snapper
Nemadactylus valenciennesi
Most commonly found around reefs, the queen snapper is related to the dusky morwong and magpie perch.

24. Blackthroat Threefin
Helcogramma decurrens
Threefin can be found 'hiding' on jetty piles, with fins twitching to attract prey.

25. Wavy Grubfish
Parapercis haackei
Found in rock pools and the ocean, grubfish will rest on their ventral fins, camouflaged against the bottom, before darting off if disturbed.

26. Pencil Weed Whiting
Siphonognathus beddomei
A cleaner fish, the females have a false eye on their tail fin as a decoy.

27. Yellowtail Scad
Trachurus novaezelandiae

28. Silverbelly
Parequula melbournensis

27

28

Leafy sea dragons

Leafy Sea Dragons (Phycodurus eques) are very elaborate fish in which the jaws have been fused into a long snout and the fins restricted in size, but converted into beautifully ornate camouflage. They are also unusual in having bony armour around their body.

They feed on tiny crustaceans and plankton from above seagrass or from sand patches and red seaweeds shaded by beds of kelp (Ecklonia) – which they closely resemble.

Their distribution is linked to suitable habitat along southern Australia and they tend to range within local patches. Not being fast swimmers, they rely on their camouflage and armour for protection from predatory fish, crabs and birds.

Unlike some sea horses, leafy sea dragons are not able to use their tail to hold onto anything and are sometimes found washed up on beaches following storms.

Males and females pair up during the early summer breeding season, and may hatch more than one brood per year. The responsibility of brooding the young rests with the males. They carry the fertilised eggs in a brood-patch along the front of their tail, as shown in the photograph below.

They are placid animals and move as if wafting along like a piece of seaweed. Their unusual beauty, and difficulty in sighting them, makes them a favourite with divers. The leafy sea dragon is the marine emblem of South Australia and is fully protected – it is illegal to take live specimens from the wild.

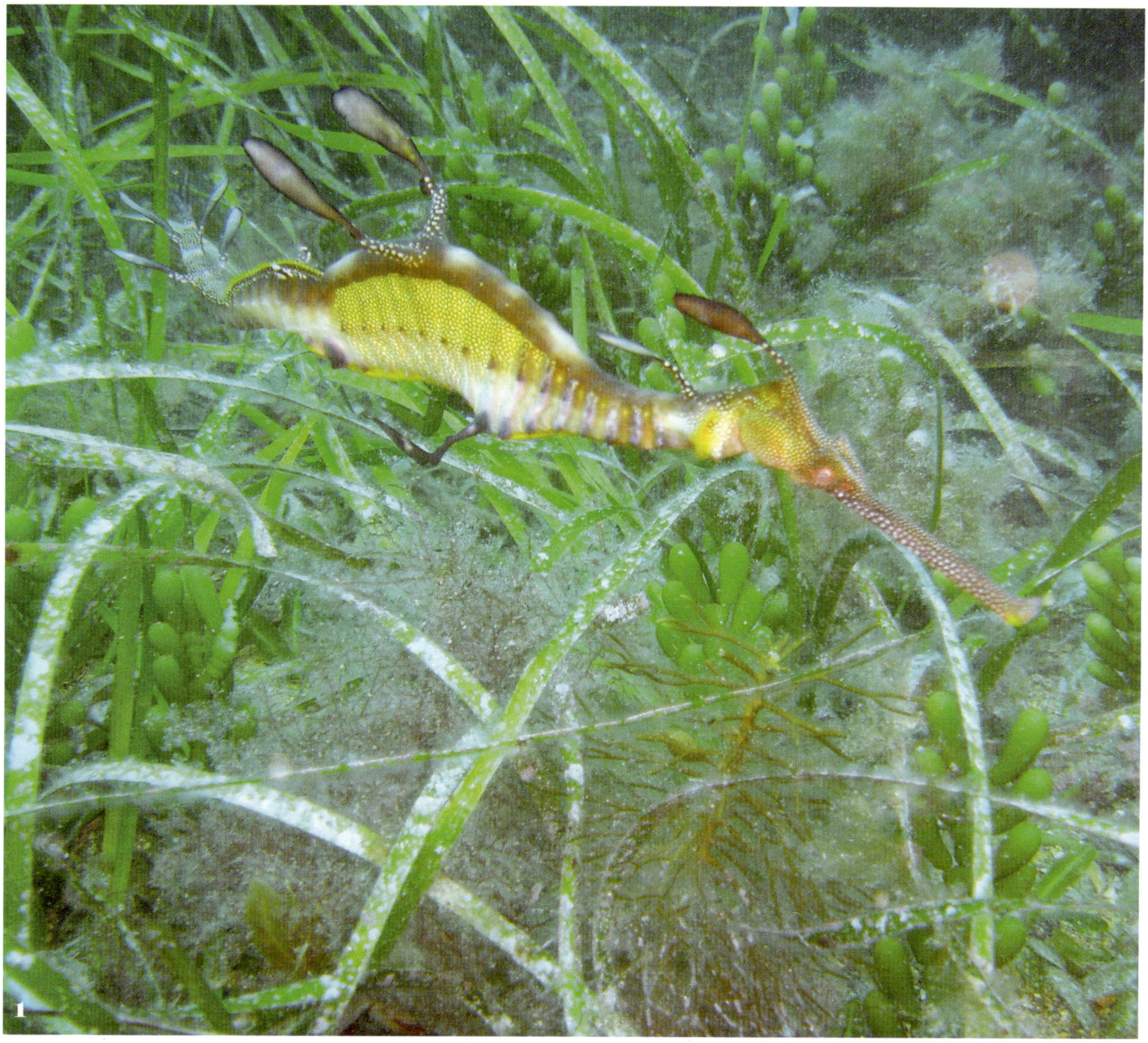

1 & 2. Weedy Sea Dragon
Phyllopteryx taeniolatus
The male in photograph 2 is carrying eggs.

Marine Mammals You Might See

1. Australian Sea Lion
Neophoca cinerea
Seals may be inquisitive and playful with divers, but can bite. Bull seals should be avoided.

Marine mammals include cetaceans (whales and dolphins – which live in the water) and pinnipeds (seals and sea lions – which can spend extensive periods in the water but retreat to dry land to rest and for breeding). All lack gills, but are oxygen breathing and must therefore surface for air. They are specially adapted to retain body heat and are streamlined swimmers. Dolphins and whales use 'echo-location' (bouncing sound off objects) to sense their surroundings. Marine mammals are fully protected in South Australian waters.

2. Bottle-nosed Dolphin
Tursiops truncatus
Dolphins will often tag along in the bow wave of boats and may be seen underwater in pursuit of cuttlefish or fish.

Australian Sea Lion

The Australian sea lion (Neophoca cinerea) *is the nation's only endemic seal and Australia's least numerous. Male sea lions are about twice the size of females and much darker in colour when mature, apart from their yellowish mane. Female sea lions may live to nearly 25 years of age while males survive until their mid-teens. When nursing their young, mothers spend half their time at sea hunting for food.*

The Australian sea lion has a breeding cycle that is unique among seals. Breeding occurs on rocky off-shore islands at roughly 18-month intervals, but each colony has its own cycle: adjacent colonies may breed at different times. The pups are nursed by their mothers until they breed again, but not all females breed each season, so some pups may be nursed through two breeding cycles, i.e. around three years.

Pup mortality is high. Around 40% of pups die before weaning due to parasites (hookworm), starvation and aggression from adult sea lions. Juveniles and adults also suffer losses. The most pressing human impact affecting the recovery of the species is entrapment in large-mesh gill nets used in shark fishing (Goldsworthy et al, 2009).

South Australia supports 85% of the world's population of Australian sea lions but their numbers are limited. In contrast, populations of the annually breeding New Zealand fur seal are booming since the historic cessation of seal hunting in local waters.

Marine birds you might see

Marine birds have some special adaptations for their watery environment. Most have virtually waterproof feathers due to waxes and fats that they secrete from glands and spread over themselves when preening. Many also have a nasal salt gland – a way to concentrate and excrete the excessive salt they take in. Some birds drop from great heights to dive into the water after food (e.g. cormorants), while others are skilled swimmers (e.g. penguins).

1. Pied Cormorant
Phalacrocorax varius
One of several species of cormorant (shags) found in the Gulfs. Unlike other water birds, cormorants do not have waterproof feathers and may be seen with wings outstretched drying off.

2. Australian Pelican
Pelecanus conspicillatus
The Australian Pelican is reported to have the biggest bill of any bird in the world.

REFERENCES

Edgar GJ (2000) 'Australian Marine Life. The plants and animals of temperate waters'. Reed New Holland. Sydney

Coleman N. (1987) 'Australian Sea Life. South of 30°S' Doubleday Australia, Sydney.

Goldsworthy SD, McKenzie J, Shaughnessy PD, McIntosh RR, Page B & Campbell R (2009) 'An update of the report: Understanding the impediments to the growth of Australian sea lion populations'. South Australian Research & Development Institute. Adelaide. SA

Connell SD & Gillanders BM (2008) 'Marine Ecology'. Oxford University Press. South Melbourne. Victoria

Gomon M, Bray D & Kuiter R (2008) 'Fishes of Australia's south coast.' Reed New Holland. Sydney.

Gowlett-Holmes K (2008) 'A field guide to the marine invertebrates of South Australia.' Notomares. Sandy Bay, Tasmania.

Huisman JM (2000) 'Marine plants of Australia.' University of WA. Nedlands. WA.

Hutchins B & Swainston R (1986) 'Sea Fishes of Southern Australia. Complete field guide for anglers and divers.' Swainston Publishing. Perth.

Kuiter RH (2008) 'A photographic guide to Sea Fishes of Australia.' New Holland Publishers, Sydney.

Kuiter RH (2009) 'Seahorses and their relatives.' Aquatic Photographics. South Australia.

Norman M & Reid A (2000) 'A guide to squid, cuttlefish and octopuses of Australia'. Gould League of Australia. Victoria.

Shepherd SA, Bryars S, Kirkegaard IR, Harbison P, Jennings JT. (2008) 'Natural History of Gulf St Vincent'. Royal Society of South Australia Inc.

Getting started

Snorkelling

The first thing to organise for happy snorkelling is a good dive mask – and 'good' means one that fits. It is no fun to have a leaking mask (goggles) or one that is uncomfortable. Hold a mask against your face, breath in and remove your hand. A good-fitting mask will stay there as a snug fit. Once you have one that suits the shape of your face then you can worry about the durability of materials, cost – and maybe colour.

To avoid having your mask fog-up underwater, rub the inside with toothpaste before first use and a gentle detergent (e.g. baby shampoo) before giving it a rinse in the sea and putting it on. As a last resort, a smear of saliva will help.

Most snorkels have a valve to reduce the risk of you getting a mouthful of water. Get one that is comfortable and practice blowing into the mouthpiece so you can dive underwater and, once you've surfaced, blow out the water that gets into the snorkel. With more experience you'll be able to begin gently blowing bubbles out the snorkel as you surface – but you need to have a good feel for where the surface is.

It is always best to snorkel with someone else – a 'dive buddy'. It's more fun and a lot safer, especially if they are a strong swimmer. When you are just starting it may be comforting to have an observer on the shore to keep an eye on you and help if needed. Sun protection for your back, neck and even legs is a good idea when you are in the water – a T-shirt or suntop will also help keep you warm. For those with thin hair, a bandana can be handy sun protection.

After being in the sea it is always a good idea to wash your gear in fresh water. Rinsing off the salt means gear stays comfortable, doesn't clog-up or get sticky, and may extend its life.

DIVING

Before anyone can go scuba diving they need to have completed accredited training – such as a PADI (Professional Association of Diving Instructors) or SSI (Scuba Schools International) course. There are dive shops in most areas of Adelaide that offer training as well as providing scuba gear. Once you have completed training your certification will be recognised anywhere around the world.

After completing training, the best advice for a novice diver is to join a dive club. Clubs provide a safe environment in which to get some dives under your belt and to build on the knowledge gained in training. They can provide the encouragement that might be needed to get diving and they have members with lots of experience who are only too happy to help and to share their favourite dive spots. Do a web search of Adelaide dive clubs to find one near you.

Clubs also often provide gear for hire, so you can buy equipment progressively instead of having to spend a lot of money at once. If you can't afford a shark-shield when starting out, try to hire one or look for a dive buddy who has one and stick close to them.

ReefWatch is another opportunity for divers whether in a dive club or not. It is a community-based reef monitoring program, helping to keep an eye on the health of the underwater environment. Volunteers are trained in monitoring techniques (a great way to learn to identify sea creatures) and organised monitoring dives provide a network of like-minded dive buddies, experienced dive leaders and a chance to do something for the environment. For more information, see www.reefwatch.asn.au or try www.reeflifesurvey.com for a national network to marine scientists.

The Marine Life Society of SA is another organisation committed to the study and protection of South Australia's underwater world. See www.mlssa.asn.au

About the Authors

Peter Day

Peter's childhood years were spent on Kangaroo Island – surrounded by sea. However, he didn't really learn to swim until given a goggle and snorkel set one Christmas. He quickly set to exploring the rock pool at Stokes Bay and any other spot he could find. Living in Adelaide during his late teens and early 20s he enjoyed snorkelling at Hallet Cove, Port Noarlunga and Second Valley. Much later in life he took up scuba diving.

Peter studied zoology and botany at university and was especially interested in marine biology and ecology. Diving soon brought back faint memories of those studies as he came face to face with the unique life forms found in the sea. Dusting off old text books and searching new references helped him identify different species and develop an appreciation of their relationships and fit with their environment – but none provided exactly what he was after.

A natural resources consultant by trade, Peter spends much of his working time interpreting and communicating environmental science. This book draws on his experience in 'knowledge harvesting'. It began as a side-project for personal interest and, with the support and encouragement of others, grew into a publication that he hopes will be of interest to many. If it helps people toward a deeper affection for the oceans and marine life – and an appreciation of their values – it will be a success for him.

Antony King

Antony's interest in diving was sparked by documentaries on marine life and a suggestion from a good friend that they should learn to scuba dive. Inevitably, he was drawn to photography to share his own diving experiences with others, his mum and grandmother especially. By methodically trying new things and seeking feedback from others, he has continually improved his photographic technique.

Antony also enjoys the social side of diving and loves to help new divers gain confidence and build upon their basic diving skills. He is a 'dive whisperer' and has guided countless novice divers over the years, helping them to overcome any initial fears and uncertainty and to gain confidence and pleasure from diving. As one of the experienced divers at Flinders University Underwater Club, Antony takes pride in seeing new members getting into diving and coming back happy after safe, enjoyable dives.

Some of Antony's images have been published by the Marine Life Society of SA and others have been used by ReefWatch in their catalogue of Marine Species of Conservation Concern. Contributing to this book is yet another avenue for him to share his enjoyment of diving with others and to illustrate how wonderful and diverse the local marine life is.

James Manna

James had been interested in scuba diving since he was a lad but it wasn't until he graduated from university that he rewarded himself with a diving course as a way to celebrate. There has been no looking back since. Within a few weeks of gaining an open water diving certificate James had completed 25 dives, with nearly half being spent hunting for crayfish; a skill he has perfected over the years.

Now, 11 years and 400 dives later, James still loves the total experience of diving: the boating, the friendships and social aspects, being 'weightless' and breathing underwater, and getting face-to-face with the amazing marine life. As he says, "It's the closest thing to being in space that I'll ever experience".

Whether diving at an exotic holiday destination or a local site he has been to dozens of times before, James enjoys seeing something new with every dive as conditions and seasons change, or as something unusual wanders into view. An experienced and skilful photographer, James loves to capture those moments and is forever looking to improve the quality of the images he takes. Contributing to this book is a way for James to share his love of diving and the marine world.

Photographic Credits

Page	No.	Photographer
Cover main image		Antony King
5	1	Peter Day
11	1	James Manna
11	2	Peter Day
11	3	Peter Day
11	4	Antony King
14	1	Antony King
15	1	Peter Day
15	2	Antony King
15	3	James Manna
16	1	James Manna
17	1	James Manna
18	1	James Manna
18	2	Antony King
18	3	Antony King
19	1	Antony King
19	2	Peter Day
21	1	Peter Day
22	1	Peter Day
22	2	Peter Day
22	3	Peter Day
23	1	Antony King
25	1	Peter Day
25	2	Peter Day
25	3	Peter Day
26	1	Peter Day
32	1	Peter Day
33	2	Peter Day
33	3	Peter Day
33	4	Peter Day
33	5	Antony King
33	6	Peter Day
34	1	Peter Day
35	2	Peter Day
35	3	Peter Day
35	4	Peter Day
36	1	Peter Day
36	2	Peter Day
37	3	James Manna
37	4	Antony King
37	5	Peter Day
37	6	Peter Day
37	7	Peter Day
37	8	Peter Day
39	1	Peter Day
39	2	Antony King
39	3	Antony King
40	4	James Manna
41	5	James Manna
41	6	Peter Day
41	7	Peter Day
41	8	James Manna
42	1	Peter Day
42	2	Peter Day
43	1	Antony King
44	2	James Manna
45	3	James Manna
45	4	James Manna
45	5	Peter Day
45	6	Peter Day
45	7	Peter Day
46	1	James Manna
46	2	James Manna
46	3	James Manna
47	1	Peter Day
47	2	James Manna
48	1	Peter Day
48	2	James Manna
49	3	Antony King
49	4	James Manna
49	5	Antony King
49	6	Antony King
49	7	Antony King
49	8	Antony King
50	1	Peter Day
50	2	Antony King
50	3	Peter Day
51	4	James Manna
51	5	Antony King
51	6	Antony King
51	7	James Manna
51	8	Antony King
51	9	Antony King
52	1	Antony King
52	2	Peter Day
53	3	James Manna
53	4	Peter Day
54	5	James Manna
54	6	Peter Day
54	7	Peter Day
54	8	Peter Day
55	1	Peter Day
56	2	Antony King
56	3	Antony King
56	4	Antony King
57	1	Antony King
58	2	James Manna
58	3	Peter Day
58	4	Antony King
58	5	James Manna
59	6	Antony King
59	7	Antony King
59	8	Antony King
60	9	Peter Day
60	10	Antony King
60	11	Peter Day
61	12	James Manna
61	13	Antony King
61	14	Peter Day
61	15	Antony King
61	16	Antony King
61	17	Peter Day
61	18	James Manna
62	19	Antony King
62	20	Antony King
63	21	Antony King
63	22	James Manna
63	23	Antony King
63	24	James Manna
64	25	James Manna
64	26	Antony King
65	27	Peter Day
65	28	Peter Day
66	1	James Manna
66	2	James Manna
66	3	Antony King
66	4	Antony King
67	1	Antony King
67	2	Antony King
68	1	James Manna
68	2	Antony King
69	1	James Manna
69	2	James Manna
70	1	Peter Day
70	2	Peter Day
72	1	Antony King

100 CREATURES YOU CAN GET TO KNOW

COMMON NAME Page

More images

For more images of local sea creatures visit:

- The Marine Index of the Marine Life Society of South Australia: www.mlssa.asn.au/marine.html
- Reef Watch manuals: www.reefwatch.asn.au/
- OzAnimals Australian Wildlife: www.ozanimals.com/australian-fish-index.html

Invasive pests

A number of invasive pest animals or seaweeds have become established in Gulf waters and some have potential to upset the local ecology. Once established they are very difficult to control so divers are encouraged to keep an eye out for invasive pests so action may be taken before they get a 'foothold'.

The 'Feral or in Peril' program run by Reef Watch has some excellent information, including brochures and identification guides. See www.reefwatch.asn.au/feralInPeril.html

The SA Museum will also help to identify any specimens thought to be a new invasive pest. A good photograph is a great way to get started.

Dive notes

Species observed	Dive site	Date

DIVE NOTES

SPECIES OBSERVED	DIVE SITE	DATE